Planets in our Solar System

Mercury

Closest to the Sun
Smallest planet in our Solar System
Orbital period: 88 days

Venus

Between Earth and Sun
Orbital period: 225 days
Rotates Clockwise

Earth is the third planet from the Sun and the only astronomical object known to harbor life.

Our planet, Earth

Mars

The "Red Planet"
4th from the Sun
2nd Smallest
About 1/2 the size of Earth

Jupiter

Fifth planet from the Sun

The largest in our Solar System

Orbital period: 12 years

Uranus

The seventh planet from the Sun
13 Rings/27 Moons
Orbit period: 84 years

Neptune

At least 5 Rings
14 Moons
Coldest Planet
Orbit Period:
164.8 years
Farthest from Moon

Venus

Which planet is larger?

Uranus

Which planet is the smallest?

Dwarf Planets

Ceres (dwarf planet)

Eres (dwarf planet)

Most massive and
second-largest known
dwarf planet in the
Solar System
Orbital period
558 years

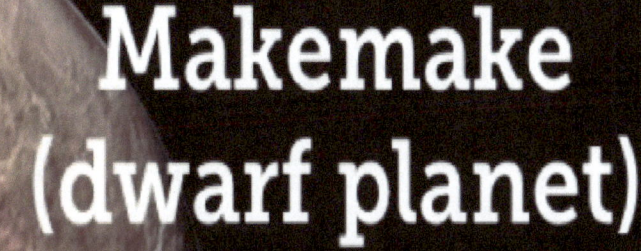

Makemake
(dwarf planet)

Orbital period:
309 years

Dwarf Planets

Ceres • Makemake • Dysnomia • Pluto • Charon • Earth • Eris

STARS

Sirius

Sirius is the brightest star in the night sky.

Rigel

Generally, the seventh-brightest star in the night sky and the brightest star in the constellation of Orion. Its brightness varies slightly, and it is occasionally outshone by Betelgeuse

Do you know the largest Star in our galaxy?

Sun

The star at the center of the Solar System. It is a nearly perfect sphere of hot plasma.

5 Major Constellations

Ursa Major
Ursa Minor
Orion
Taurus
Gemini

What is the name of our galaxy?

The Milky Way

The galaxy that contains our Solar System, with the name describing the galaxy's appearance from Earth: a hazy band of light seen in the night sky formed from stars that cannot be individually distinguished by the naked eye.

Thank YOU for Learning about the Universe!

Definitions

Dwarf Planet: A planetary-mass object that does not dominate its region of space and is not a satellite.

Orbital Period: The time a given astronomical object takes to complete one orbit around the Sun, moons orbiting planets.

www.ingramcontent.com/pod-product-compliance
Lightning Source LLC
Chambersburg PA
CBHW051940210526
45473CB00006B/2319